THE SCIENCE OF ANIMAL MOVEMENT

How Critters Climb

BY EMMA HUDDLESTON

CONTENT CONSULTANT
DAVID HU, PHD
PROFESSOR
MECHANICAL ENGINEERING
GEORGIA TECH

Kids Core
An Imprint of Abdo Publishing
abdobooks.com

abdobooks.com

Published by Abdo Publishing, a division of ABDO, PO Box 398166, Minneapolis, Minnesota 55439. Copyright © 2021 by Abdo Consulting Group, Inc. International copyrights reserved in all countries. No part of this book may be reproduced in any form without written permission from the publisher. Kids Core™ is a trademark and logo of Abdo Publishing.

Printed in the United States of America, North Mankato, Minnesota
042020
092020

Cover Photo: Hadrani Hasan/Shutterstock Images
Interior Photos: Shutterstock Images, 4–5, 10, 26; Tom Meaker/iStockphoto, 6; iStockphoto, 7, 16, 28 (top), 28 (bottom), 29 (top), 29 (bottom); Don Mammoser/Shutterstock Images, 8; Alex Popov/Shutterstock Images, 12–13; Ted Kinsman/Science Source, 15 (top left); Cheryl Power/Science Source, 15 (top right); Federico Crovetto/Shutterstock Images, 15 (bottom); Charles Siu/Shutterstock Images, 17; JBStudios/Shutterstock Images, 18; Steve Gschmeissner/Science Source, 20; Adrian Coleman/iStockphoto, 22–23; Linas Toleikis/iStockphoto, 24

Editor: Marie Pearson
Series Designer: Ryan Gale

Library of Congress Control Number: 2019954245

Publisher's Cataloging-in-Publication Data

Names: Huddleston, Emma, author.
Title: How critters climb / by Emma Huddleston
Description: Minneapolis, Minnesota : Abdo Publishing, 2021 | Series: The science of animal movement | Includes online resources and index.
Identifiers: ISBN 9781532192944 (lib. bdg.) | ISBN 9781644944332 (pbk.) | ISBN 9781098210847 (ebook)
Subjects: LCSH: Children's questions and answers--Juvenile literature. | Animal climbing--Juvenile literature. | Science--Examinations, questions, etc--Juvenile literature. | Habits and behavior--Juvenile literature.
Classification: DDC 500--dc23

CONTENTS

CHAPTER 1
Finding Grip 4

CHAPTER 2
Dry Grip 12

CHAPTER 3
Wet Grip 22

Movement Diagram 28
Glossary 30
Online Resources 31
Learn More 31
Index 32
About the Author 32

A squirrel uses its sharp claws to climb.

CHAPTER 1

Finding Grip

The forest's leaves are bright red, orange, and yellow. A squirrel is busy finding acorns to store for winter. It climbs up a tree. Its sharp claws dig into the bark. It pulls its body up.

A squirrel's ankles rotate to allow its claws to hook onto bark as it climbs down a tree.

Then the squirrel climbs down the tree headfirst. The claws on its back feet can still cling to the bark. That is because a squirrel can **rotate** its ankles backward. This lets the squirrel's claws also point backward. They can hook onto the bark. The critter's short **limbs** keep its body close to the tree. The squirrel can better control its movement.

A cat's claws help it climb trees.

Fighting Gravity

Many animals, such as birds and lizards, use claws to climb. Claws are hard and pointed. They can pierce into wood or cracks in rocks. This gives animals grip. With grip, the animals can climb. A squirrel's claws hook into grooves on tree bark. The squirrel can unhook its claws to move its foot.

Monkeys have fingers that can grasp branches.

Large animals have to move their heavy bodies against gravity to climb. Gravity is a force that pulls objects toward Earth's center. Monkeys are large climbing animals. They grab branches above and pull their bodies up. Their muscles pull them up against gravity.

Other animals such as insects and reptiles often have smaller bodies. They don't weigh as much. But they still have to fight gravity to climb. And the surfaces they climb are often fairly smooth. Their feet can't grab these surfaces. They need other ways to climb.

What about Snakes?

A snake uses friction to get around. It doesn't have limbs for walking or climbing. Instead, scales on its belly rub against the tree bark. The scales grip the bark so the snake can push off and move upward.

Many lizards are great at climbing trees.

Making Friction

Claws are just one tool that helps critters climb. There are other ways critters apply forces to a surface. When two surfaces are pressed together, they create a force called friction. Friction resists slipping. Friction helps critters climb. Claws use friction. So do other body parts.

Many climbing critters have body parts that increase friction. Soft skin and bumps on toes are important. When an animal touches a surface, soft skin bends along with any bumps on that surface. Some critters have especially amazing ways of climbing. Science explains how this works.

Explore Online

Visit the website below. What new information did you learn about reptiles that wasn't in Chapter One?

Reptile Pictures & Facts

abdocorelibrary.com/how-critters -climb

Geckos kept in captivity sometimes crawl on the glass sides of their enclosures.

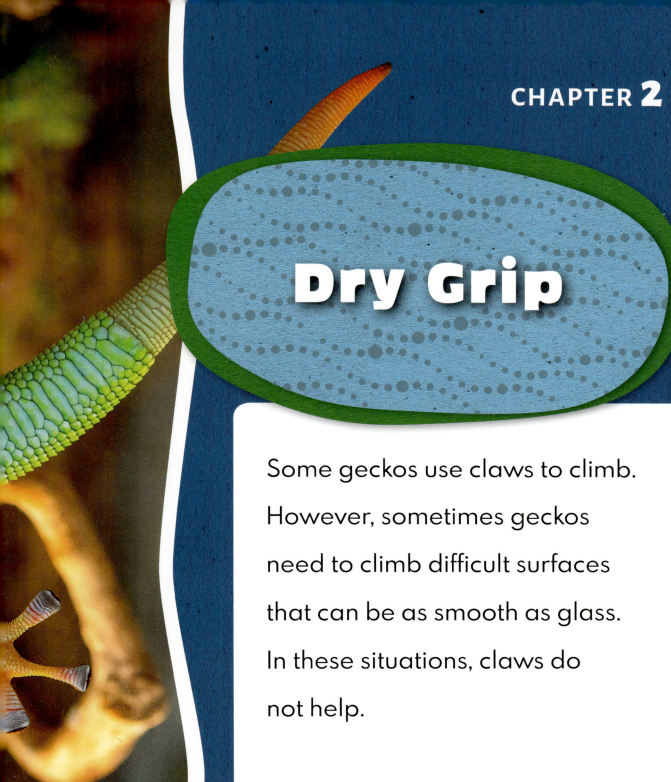

CHAPTER 2

Dry Grip

Some geckos use claws to climb. However, sometimes geckos need to climb difficult surfaces that can be as smooth as glass. In these situations, claws do not help.

Claws are too big and firm to fit in the grooves of smooth surfaces. It may seem like geckos' feet must be sticky to climb up glass. But geckos actually rely on friction while climbing these surfaces.

Tiny Hairs

Even smooth surfaces have tiny cracks and bumps. These cracks and bumps are too small to see. But they make it possible for geckos to find grip. Geckos have tiny plates on the undersides of their toes. These plates are called lamellae. The lamellae have hundreds of tiny hairs called setae. The setae branch out at the tips into even tinier **bristles**. The bristles are called spatulae.

A Gecko's Foot

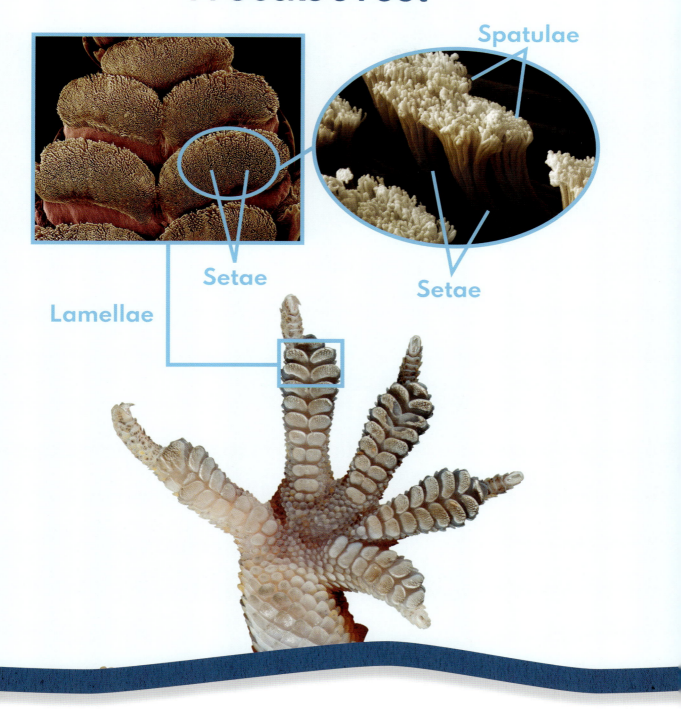

A gecko's foot has special climbing parts. The lamellae, setae, and spatulae play an important role in helping geckos climb.

Geckos can climb walls with their amazing feet.

A gecko uses the tiny hairs to climb. The hairs allow the gecko to use van der Waals forces to stick to the wall. These forces cause very close surfaces to be pulled toward each other. Because the spatulae are so tiny, they can get very close to the climbing surface. The hairs get so close to the surface that they get pulled to it.

Geckos can hang upside down.

Like geckos, anoles have toe pads that help them climb.

The setae grow angled toward the end of the toe. When a gecko presses its feet on a surface and drags forward a bit, the hairs catch on to the surface. They attach so strongly that they can hold the gecko's body weight easily. Geckos uncurl their pads to release grip.

Other Climbers

Geckos aren't the only climbers that use tiny hairs. Like geckos, lizards called anoles can climb smooth surfaces. Their feet have similar tiny hairs.

Life in the City

Anoles in Puerto Rico live in forests and cities. Cities have smoother surfaces than forests. City anoles have adapted to the different habitat. They have more lamellae to help them climb walls and windows.

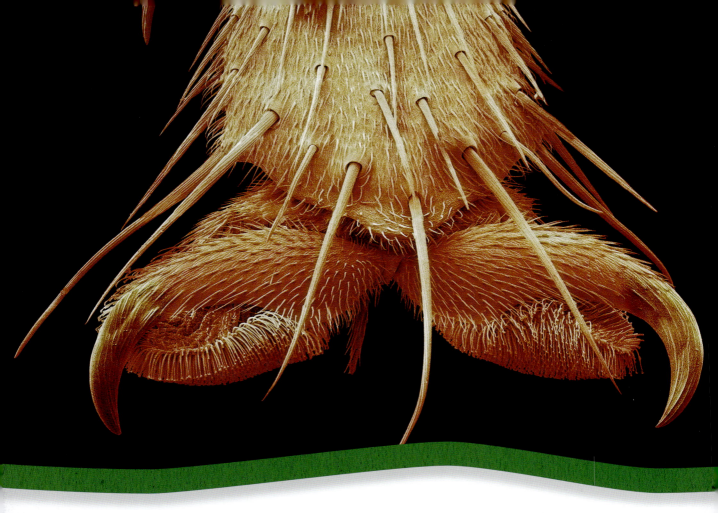

A fly's foot has many hairs that help it climb.

Insects often climb waxy or smooth surfaces such as leaves. Many insects have tiny hairs on their legs, similar to those on a gecko's feet. Other insects, such as ants, produce an oily substance on the tips of their legs. The substance is sticky. It helps them attach to surfaces.

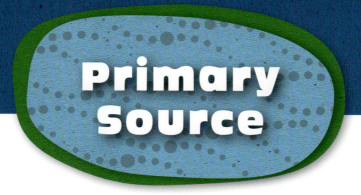
Primary Source

Alex Greaney has studied how a gecko's foot works. He said:

> A gecko by definition is not sticky—he has to do something to make himself sticky. . . . It's . . . the [incredible] hairs that [make] it possible.

Source: Kelly Dickerson. "Geckos' Sticky Secret? They Hang by Toe Hairs." *Live Science*, 12 Aug. 2014, livescience.com. Accessed 16 Jan. 2019.

Comparing Texts

Think about the quote. Does it support the information in this chapter? Or does it give a different perspective? Explain how in a few sentences.

Tree frogs can wrap their feet to grip a surface as well as use special toe pads to climb.

CHAPTER 3

Wet Grip

Amphibians such as frogs need wet habitats to survive. Water on the ground and in the air helps keep their skin from drying out. Some frogs are great climbers. Since frogs live in wet places, they have to be able to climb on wet surfaces.

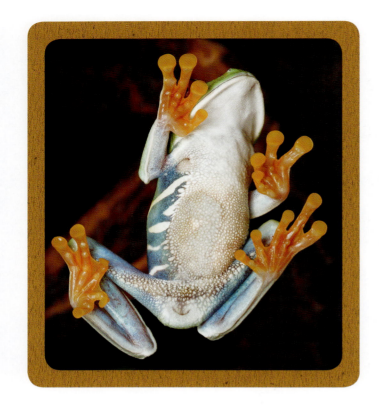

Tree frogs have toe pads that help the frogs climb glass.

Tree frogs' bendy toes and limbs help them find grip. They grasp twigs and other climbing objects. The frogs can also climb smooth, flat surfaces such as glass. Friction and stickiness help them climb easily.

Layer of Mucus

Pads on tree frogs' toes are divided into small sections. The sections are **hexagon** shaped.

Each section has many tiny bumps. In the cracks between the sections is mucus. Mucus is a gooey substance that sticks to some surfaces. Since only small amounts of it fill the space between sections, there isn't so much mucus that it separates the frog from the surface it is trying to stick to. Just the right amount of mucus helps increase friction.

Clean Feet Help with Grip

Tree frogs can clean their feet between each step. They make more mucus. They rub their feet on the ground when they push off to take a step. That movement helps leave dirty mucus behind.

Tree frogs can climb wet surfaces.

The bumps and grooves of a tree frog's foot act like a tire on a wet road. If the surface is wet, water gets pressed between the cracks in the toe pads. At the same time, the frog pushes its toe pads against the surface. With the water pushed away, the bumps can get very close to the surface. This pressure creates friction.

Critters climb by getting good grip. Some use claws or tiny hairs to hook into cracks and bumps. Others rely on close contact to keep from slipping. All are amazing climbers.

Further Evidence

Look at the website below. Does it give any new evidence to support Chapter Three?

Friction

abdocorelibrary.com/how-critters-climb

Movement Diagram

Tree Frog

Footpads and mucus help it climb wet surfaces.

Gecko

Tiny hairs on the toes help it climb walls.

Glossary

adapted
changed as a species in order to better survive

bristles
hairs that are short and stiff

habitat
a place where something naturally lives

hexagon
a shape with six sides

limbs
body parts such as arms and legs that branch out from the main body

rotate
to turn in a circle around a central point

Online Resources

To learn more about how critters climb, visit our free resource websites below.

Visit **abdocorelibrary.com** or scan this QR code for free Common Core resources for teachers and students, including vetted activities, multimedia, and booklinks, for deeper subject comprehension.

Visit **abdobooklinks.com** or scan this QR code for free additional online weblinks for further learning. These links are routinely monitored and updated to provide the most current information available.

Learn More

Howell, Catherine Herbert. *Reptiles & Amphibians*. National Geographic, 2016.

Huber, Raymond. *Gecko*. Candlewick, 2019.

Index

anoles, 19

claws, 5–7, 10–11, 13–14, 27

friction, 9, 10–11, 14,
  24–25, 27
frogs, 23–27

geckos, 13–20, 21
gravity, 7–9
Greaney, Alex, 21
grip, 7, 9, 14, 19, 24, 27

insects, 9, 20

lamellae, 14, 15, 19

monkeys, 8–9
mucus, 25

setae, 14, 15, 18
skin, 11, 23
snakes, 9
spatulae, 14–16
squirrels, 5–7

toes, 11, 14, 18, 24, 26

van der Waals forces, 16

About the Author

Emma Huddleston lives in the Twin Cities with her husband. She enjoys reading, writing, and swing dancing. She thinks the science of animal movement is fascinating!